My pet clock ticks all day. It sounds grossly disturbing but it isn't. Like you may have a string

tied to the clock and slowly pull the clock around though as not to tip it over. You remind yourself

freely "don't hurt the pet clock." It is black and white like that of a chess board.

My Little Pet Clock.

You linger in my mind like my dad. Dwindling. Why couldn't he finish a sentence in his old age I don't

know.

It seems weird. The days passed on and on and I still have this pet clock. It sits. Sometime's I

feel like it's talking to me. It says, " Hey do you know what time it is?" "Because I'm going to be late

for my date." A clock having a date with another clock? I say, how does that work?

I see other things such as the deodorant, earrings and neclaces but none keep my interest as

such-my pet clock.

It's hard to say when the day will come that we see the clocks turned back. I often wonder about things

like time travel.

I wonder if you turn the clock back, for example from 9:15 am to 9:00 am then do you for that

moment, that instance go back in time. Yet I wonder if that's too easy. On days that I've run into the

Marriot Hotel and it's sunny out, I feel time travel too.

So do you ever feel like your the only one wondering about time travel? Like when there's a

room full of persons in a hall at a college and you or me are looking up, taking in the amount of space

and for that one split second you feel love. Love could be time travel. You look up and you see test

takers everywhere. And I mean everywhere. (WINK) ah nevermind, thats the end...

Update: like a phone is my policeman?

Update: money, car, house all a means to murder people

Update: thought they tricked me into thinking jealousy is about looks infact it is about money- now I just like to say I was suicidal because I had reminints of carpul tunnel syndrome-hurts like a mother, much less actually

Update: I guess you wanted it that way all along- he gave me no frites, no bread, no sausagay- how can one function when one is oogroo for ʌ, and a little bit of lava. Oogroo for lava. If only I had a little neeblet- point taken. GET OUT >? Is anyone up for adoption>? I said shut up. Como!!!!!!!!!!!!!!!!!!!!!!

Update: goolash is good for the heart. The heart and soul baby, it's the heart of the matter but my will gets weak and my thoughts seem to scatter but I think it's about forgiveness, forgiveness, even if, even if, you don't love me anymore.

Update: Dubbed: In those times of need
Undubbed: We weep, we love, we forget.
About marilyn monroe. I can relate. I is Mirjana Nikolovski.

Update: to be or not to be that is the question!

Update: cricket to cricket just pulled me out of school came from the edge of town a switchblade motor running but you better not take it from bang bang shoot em dead but the party never ends, u will see u but the but the par par never ends stick it to ya will see the party in the party in the switchblade you better not take it from me -skidrow and me mirjana is that ok? Wa wa wa

Update: Story entitled "Computerized Image"

Before you walked into my life, I waited for you and new that you had existed.

Although you are not real, you are only a computerized image.

You are the best so far.

Has there ever been such a power, to bring human and non together, or is it just a dream.

This is the dream.

Which has happened not to be.

I can't tell, for I am human and you are not.

But eternity is only an obsession, between all humanity.

Through us there is no eternity, only expiramental viewings and love.

Not love, but love impaired.

Our creation is becoming weaker and weaker, as yours is becoming stronger and stronger.

But can you help me come into your world, or will you let me the creator of your kind, wilt away and die.

It is your choice, but there shall be no garentee, that you will make it through, the power of this world.

So choose soon, because I can hear you, but you can't hear me, for the power is making your voice weaker and weaker, as mine is getting louder and louder.

Update: Don't cry

Don't cry because someone once told me it was scary. I still love you a lot even if we don't see eachother that much. I'll always love you. Sometimes I ask myself, are you thinking about me when I'm thinking about you, and sometimes I wonder when I'm crying for you, are you crying for me. I remember that one night, when I first met you and how much I love you, and I still love you as much as much as I loved you the first time we met eyes, and I always will, but now I have to tell you some news your heart will prabably refuse. but it's for the better. I or we can't go on this way, fore I am experiencing way to much pain, for you? A kind of pain I have never felt, I miss you to much? I love you to much, for you I just feel to much. Perhaps this all would have come out different if we had been together from the beginning. But we are not so for now who knew maybe forever I must say goodbye and good luck to you. I love you, don't forget that? But don't bother writing me, because I prabably won't right back. But never forget the memory. The memory. Try to rememeber. Try to remember the memory of our once being together. I love you. Only you, you.

Update: Eulogy

Me and my loved one

time and time again I see a glimpse of glimmering light,
time and time
to the president of the united states of america
my mother and sister are trying to kill me can you please help me
eulogy, just because someone is trying to kill you kill she said something must be done lol
lolololololololololololololololol

Update: Investigation on my dad's murder

I had my hair pulled by sister and then my dad fell down the stairs. Someone pushed him. In self defense I pulled her hair. My dad strangely died. Maybe he commited suicide because of how annoying my mom and sister are. I might. I did. I am still alive today.

Update: To the whole plaenet: Go to www.chloeandisabel.com/boutique/mirjana for beautiful jewelery. It is New York. There are bracelets, diamond bracelets and beautiful garments too.

What really is important to me is to live the happiness I love which means to live the dreams and dream the dreams. If not for these things then what else is there? I think to love is to dream. Thank you for this kind opporitunity and for helping me get my start. Thank you for trusting in my writing and I trust that you will help me in my writing pursuit. Thank you kindly. It is greatly appreciated. Title: Honey, The Appetizer I love: Sometimes I tink I have too much honey. I find myself wanting more and more of it as the days go on. Why? Well, I can not answer that out I do know that without honey my soul is drained. A drained soul so to speak. It all started when I fiound myself thinking, "I need honey." That is when I decided I had to walk to the kitchen and get it. Green Tea-that is usually the drink that calls out honey's name. Green Tea: "Honey!" Honey: "Yes dear." Green Tea: I need you...I'm a little bland today." (Ba dum Ching) and thats how it goes. Honey is like an elixer. Not to sweet like sugar cane and not to strong like mildew, but right in the middle, like honey, ah...ahhh... So I walked to the kitchen. Had a sip like I normally do. And thats when it dawned on me. This honey tastes so good. This is (for lack of better words) AMAZING! This honey is really somethin to love. This honey warms my throat. This honey warms my heart indeed. So I drank my tea. "Honey," mother calls. "Which one?" I say. "Oh Sherice, such a kidder. What will I ever do with you?" "I dunno mummy, send me away?,," I said. "Oh Sherice, my little twylark," and thats how my mornings usually go. Later that day, I found myself walking to a candy shop. On my way, I thought about things but not just any things.All sorts of thing. I thought about life. You know, the future. But most importantly, what would I buy at this candy shop? I thought and I thought gumballs, liqorice and bubble gum! Gumballs, bubblegum-same thing but my oh my bubble gum just sounds so much more appealing. Mmmm...bubblegum. Suddenly, as I walked I came to a sign-"FREE HONEY". I dropped all my bags and ran.I began to run and then decided to slow down.I wouldn't want anyone to think I was to excited. "Keep it cool...you," I said again. So I walked toward the door. Chin up, collar fixed. I decided to pull out a comb. Give my hair a quick touch up. You know, give it that cool sleak look. I open the door and walk in. I look inside and walk toward the counter. The moment I didn't know I was waiting for, all day. I look at all the people behind all the counters. They look kinda scary. So I walk in. "S'cuse me ma'am." I said the sign says FREE HONEY." Don't you shout at me young lady. "Shout?" I thought. "Lady?" I thought. I was simply expressing my happiness for Free Honey! I can't ever keep calm in my thoughts! "Sheesh-what a world. What a wonderful world...I see trees of green...red roses..." Ahh...tangents. Can't live with em...Can't live with em. "Ma'am I was shouting. To put it simply, I would just like some honey..." "To put it simply we just gave away the last drop of honey." "Oh" I said. " Well, do you know where else I could get some? " " I suggest you go to the honey shop across the street." So within the next minute I found myself walking to what I thought would be the place that would give me some honey. I open the door and walk in. "S'cuze me sir. I need some honey". "I'm sorry sir we just gave awat the last quatro but we do have the special sampling honey if you like,"he said. "Sure, anything." Well, alright. Seems your patient enough.
"Where abouts are you from?" "Just down the street sir. "Its a long way for me but I do it for the honey." "Alright young man. Now don't take it in to fast. Honey is like cool breeze on a summer day. You have to take it in slowly...otherwise you might just miss out on the taste." "Hm...OK." I said. So the man gave me some honey. Now all I had to do was find a place to eat it. I thought of all the great people I would eat it with. I bet philosophers love horey. I don't know much about them. My dad told me about them once and I didn't undetstand a word of what he said. It was to much for me. He's smart. All I know is people think I'm clever when I say the word philosopher. "Mildew, no, no one likes that word, and so I did get my honey and I did decide to just eat it. I thought of all the people I could eat it with, Tarzan, head hanchoes of clubs and other various people,and I decided to eat it alone because well, I really couldn't think of anyone who would get here fast enough to eat it with me. "Boy I love honey" and that was the end of my day.
 :| BOW

Update: Riddle me this, riddle me that, whose afraid cf the big black bat.

Update: "Ooo, poor Toto, you've come back."-The Wiz/The Wizard of OZ (OZ, why wasn't I put in that movie? Because there was air condition this time? And so it goes? Woof.

Update: You see I was infact famous in Chicago. Believe it or not no one recognized me except my dad! He funded everything. He was like here is 10,000 dollars off to Chicago you go to be famous. He is famous to but back to meeeee. I stayed in fancy hotels like the White Hall Hotel and I took cabs to The Chicago Actors Studio with flashing cameras everywhere. I got the part at The Chicago Actors Studio as the hooker without lines. Denis Fogel is nice and so were all the other actors. We had pizza and pop too! Yippy ya yo ka yay! Lets all ride onto a donkey and into the night..

Update: Then after The Chicago Actors Studio, or another time I can't remember I ended up in Memphis (my birthplace) and stayed in this cozy hotel a place that looked like Elvis would stay in. Like a Cottage Inn. Funded by my dad again! Aw papa!

Update: I performed Acting, a monologue called Somewhere for the Community House, The MSU School of Music in Detroit and at The First Prebyterian Church in Birmingham, Michigan. It goes a little something like this.

Somewhere by Mirjana Nikolovski
In a city far off,
In the land of grey,
There was a day,
That kept me through the night,
To feel soothing sweet dreams,
Where time split seams,
To see crows dancing in the air,
To be caught by the skies eyes,
To be caught by chances a night,
There is time to some,
There is a time to be,
In a free of lasting somethings,
There's always
A time to be
Somewhere

Update: Do you like the cd the big ones?

Update: Flowers in the attic

Update: Don't make me come after all of you kitties. Especially you Jimmy. Please help me get to your NY!!!!!!!!! And NPR ofcourse, ahem!

Update: Jimmy! I have so many interviews I can't figure out where I am going next to all my dyslexia. Please help. Will work for food!!! !!!!!!!!!!!!!!!!!!!!!!!!